# Millers' Tales
## -THE MILLS OF YORK COUNTY-

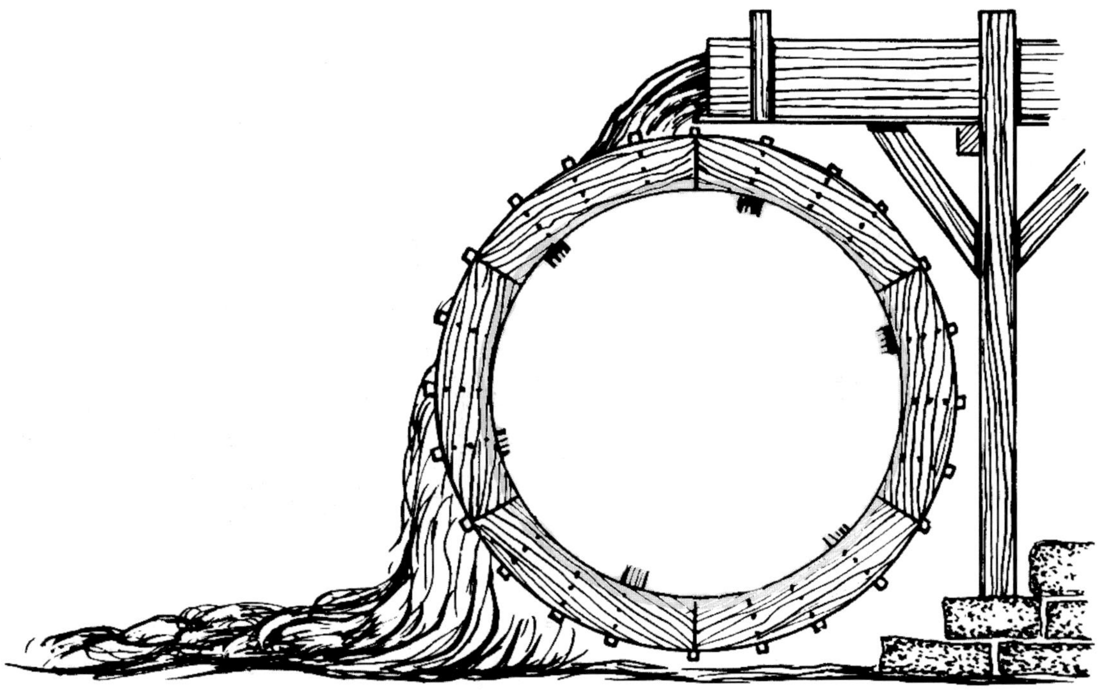

Lila Fourhman-Shaull

York County Heritage Trust
York, Pennsylvania

YORK COUNTY
HERITAGE TRUST

This publication has been produced using funds from
The Robert P. Turner Publication Fund.
The York County Heritage Trust's Publications Committee
coordinated this project.

Publications Committee
*James McClure
Philip Avillo
Lila Fourhman-Shaull
Gerald Hartdagen
Ron Hershner
*June Lloyd
John Noone
Marjorie Marsh
Joan Mummert
Patrick Rooney

*Editing Committee
Copy Editor: Joseph McClure
Indexer: Victoria Miller



ISBN



Printed by Ream Printing Co., Inc.
York, PA

This book may be purchased from:

York County Heritage Trust
250 East Market Street
York, Pennsylvania 17403

www.yorkheritage.org
Phone: 717-848-1587
Fax: 717-812-1204

Credits:
Front cover: The Wallace-Cross Mill (see Page 125) operates today as part of the York County Parks System. Cover Design; Samantha Dellinger, York Daily Record/Sunday News.
Photo: Paul Kuehnel, York Daily Record/Sunday News.
Inside title page: Waterwheel drawing from "Old Mills in York County"
Back cover: Bradley Mill is a re-constructed working gristmill located at the Agricultural and Industrial Museum of the York County Heritage Trust.
Photos: Ryan Hayman, Hayman Studio of Commercial Photography, Inc.

# Foreword

# Millers' Tales

Here begins a sampling of photographs compiled from two collections in the York County Heritage Trust Library & Archives: the Grant Voaden Mill Collection and the photograph albums of G. Ross Bond.

Both collections feature photographs of York County mills, but only Voaden's is accompanied by files of research. It may be noted that the two collections differed in names chosen for some of the York County mills. Discrepancies were also found where one image was used for two different mills, and library volunteer Albert Rose spent many hours comparing images from these collections. The following images were chosen from both of these collections due to their quality as well the availability of reliable information. An alphabetical listing of the property owners and products as compiled by Voaden are included with each image, but neither listing should be considered complete.

The files compiled by Voaden reflect the mill name that he had assigned. The contents of these files vary in depth of research and types of material. Information includes abstractions from deeds, tax records and wills, as well as interviews with owners. Some mill files contain a very detailed lineage of the property owners while the next file shows the need for additional research. These owners' names are listed alphabetically. The spelling of the name of each mill can vary depending upon the source. Bare's Mill is an excellent example as some records identify its name as being spelled "Bare," but other records spell the name as Bair, Bear and Baer. Surnames of owners can also be spelled differently from one generation to the next. And in some cases, only the last name was found within the research.

Using the 2002 map *Major Streams in York County, Pennsylvania* through the courtesy of the York County Planning Commission as a guide, 130 photographs are organized by the stream upon which they were located. There is a listing for Bennett Creek, which is featured on this map but not identified. It also should be noted that within York County several different streams share a common name. Beaver Creek and Fishing Creek are two examples. Each image is identified with the Voaden Mill number and name (GV #13—Miller's Mill). Information about the names of the York County streams that were used for the organization of this volume can be attributed to H. C. Frey, an authority on local Indian history from the 1930s.

It is the York County Heritage Trust's pleasure to share these images.

Joan Mummert
President/CEO

## INTRODUCTION

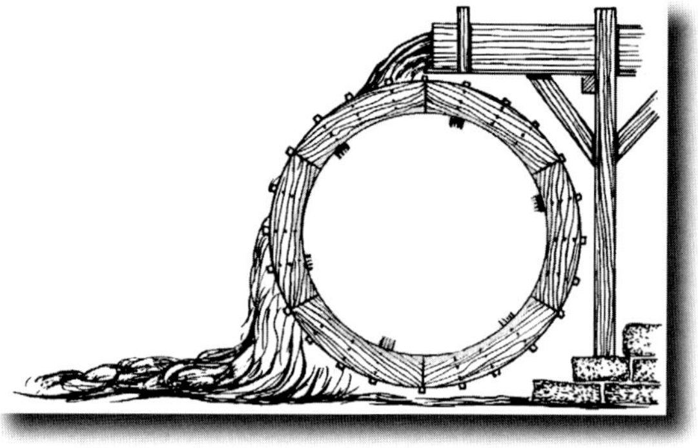

Don't let all those green acres of farmland fool you. York County has deep industrial roots.

Forestry, mining and quarrying are examples of primary industries in which raw materials are extracted from the land. The goods produced from these raw materials are examples of a secondary industry. Milling, an excellent example of a secondary industry, was one of the first industries found in York County, and for many years was its No. 1 enterprise. Abstracts from the petitions for roads found with the pages of Gibson's *History of York County* contain numerous references to mills several prior to the formation of York County in 1749.

A petition was filed in Lancaster County court in November 1741 for a road from the new town on the Great Codorus (York) to William Smith's Maryland-patented land some 10 miles south. The report, filed the following February, described the road's location: extending "due south by way of Woolrich (Ulrich) Whisler's mill." References found within several York County deed books report that "the Proprietaries of Pennsylvania by license of Samuel Blunston dated December 9, 1734, granted to Woolrich Weesler 250 acres on Little Codorus Creek about a mile above Thomas Perrens to be taken on both sides of said creek for the convenience of building a mill." Grant H. Voaden's research identifies this as the Plank Road Mill. As York County's population grew, so did the need for improved roads. Many petitions for such roads used two primary businesses as their beginning and ending points: taverns and mills. Both supplied not only benefits for business but also a location for social activities. The tavern served not only a meal but gave a place to stay and also a site for social activity—so, too, did the mill. The mill office was filled with customers as well as neighbors discussing current events and community activities.

The location of a mill required a reliable source of moving water, and York County has numerous suitable creeks and tributaries. If the flow in these waterways was not swift enough to power waterwheels, resourceful York countians built dams and races to create the required energy. These waterwheels gave rise to other York County industries and related businesses. They included the companies that built turbines and other equipment for the mill, as well as for farming and lumberyards. Mills and milling tell the story of people—our forefathers—and provide great insight about their purposeful lives, their ingenuity, their industry, their thrift, and their courage to go on despite floods, fires, terrible accidents, and sometimes bankruptcy.

+++

S. Morgan Smith and Company was named after its founder, Stephen Morgan Smith. Smith, a Moravian minister during the 1860s in York and in Canal Dover, Ohio, was forced to relinquish his profession due to a throat ailment. Inventing one of the first working washing machines, he received a patent and began to manufacture them in local York shops. Investing in York Manufacturing Company, which operated a small foundry and machine shop that manufactured waterwheels and farm machinery, proved unsuccessful for Smith. As the years passed, he gained mechanical and business experience as well as personal contacts with local millers. Then Smith, with the aid of Jesse Krall of Kralltown, invented a hydraulic turbine: the "Success."

Amos C. Jacobs, owner of a gristmill on Beaver Creek, showed interest in this turbine, and upon learning the "Success" turbine could be patented if it could show practical use, he ordered an 18-inch turbine in August 1877. A wooden penstock was built, installed and proved successful, leading to the granting of the patent. Its use extended from gristmills to other industries such as textile mills, pulp and paper, as well as lumber mills. In 1890, a new 50-by-150-foot factory employing about 20 mechanics was built. Plant production began on May 10, 1890, and it was at this time that two industries joined: the manufacturing of water turbines and electrical generators. In 1898, Smith organized the S. Morgan Smith Company, and management was turned over to his sons.

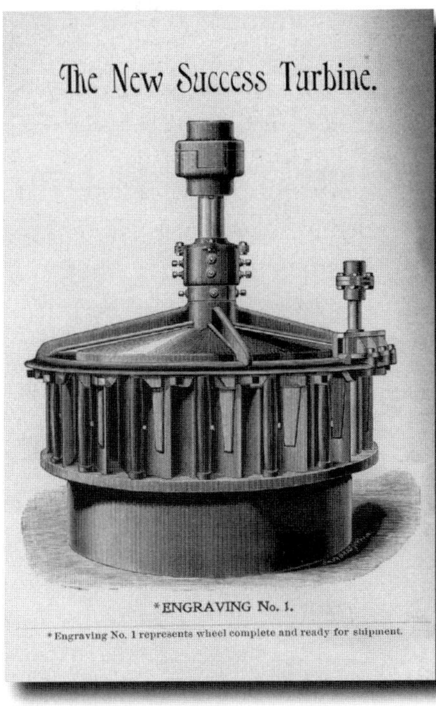


The late Grant H. Voaden, S. Morgan Smith engineer and one of the editors of the company's *Turbine Topics* magazine, spent nearly 40 years researching and photographing York County mills. This work began as a hobby and grew into his passion. The Ontario, Canada, native, arrived in York in 1925, following the guidance of a college professor. He began in the drafting room at S. Morgan Smith. There, he worked until his retirement in 1968. This included his time spent as the chief test engineer on the Smith-Putnam wind turbine. Voaden was involved in the design of that machine, and it would serve as a subject of his first article for *Turbine Topics*. The purpose of this magazine, as described by Beauchamp E. Smith, president of S. Morgan Smith, was to "develop company spirit, pride in workmanship and in product, better acquaintance with each other and fuller understanding between shop and office."

Voaden wrote in a 1977 letter the following: "In regard to my personal interest in mills, the enclosed blank form is what I am attempting to fill out for each of the approximately 360 mills which are shown on the township maps in Pomeroy's *Atlas of York County* published in 1876. I plan on writing a book on this subject and have made considerable progress on it."

This template or form that he created was to organize and guide his research for each mill. The form, titled *Index of Old Mills in York County, PA*, began with the name of the mill popular at the time of his research. It includes the names and lineage of previous owners, date of construction, type of mill, stream location, and notes made during interviews with individuals. His research also includes information from county records such as deeds and taxes, as well as maps and newspaper articles.

Below are excerpts from his third article "Old Mills and Turbines of York County" published in the October 1945 issue of *Turbine Topics.*

*"The first mill in York County was built on Kreutz Creek about 1735," according to* Spangler's Annals *written in 1896 by Edward W. Spangler of York. "The first mill in the western part of the county was in Paradise Township and was erected about 1750." This mill has*

*remained in the Jacobs family ever since, and S. Morgan Smith sold one of its descendants, George B. Jacobs, an 18" Success turbine July 15, 1878.*

*Northwest of York, the earliest mill was erected by Martin Weigle about 1738 on the Little Conewago near the present road between York and Dover, but on the opposite side from the present building known as Katz Mill. According to Gibson's* History of York County, *Martin Weigle had first tried to build a mill on the Codorus but had found the undertaking too great. He then transferred to the Little Conewago, a much smaller stream, which could be more easily dammed. While working on his mill one day, he allegedly discovered that a party of Indians was watching him from the bushes. Judging that they were full of curiosity, rather than animosity, he took a good drink of cider from a jug in full view of the Indians and proceeded to sing and dance as he worked, whereupon the Indians came out of the bushes and wanted to play too. He gave them each a drink, and that day the entire mill race was dug.*

*Prior to the construction of these mills, the settlers had to take their grain across the Susquehanna to Lancaster County to be ground, although those with only small quantities often ground it themselves in a "pioneer mill," a mortar hollowed in the end of a log or stump in which it was ground Indian style with a pestle. Near the Harmony Grove Church, there was a large mortar for grinding corn hewn out of large rock. This was probably made and used by the Indians—the Conoys and the Shawnees—who had inhabited the county, especially along the Conewago.*

*The early mills were constructed of material available at the site, mainly wood. There were many types of wheels – for example, current wheels, flutter wheels, undershot, tub, overshot, and breast wheels. The current wheels were just mounted in a stream where the current was swift, and very little, if anything, was done to the waterway. They were very inefficient. For the flutter wheel, the waterway was shaped or built up in the form of a curve, the circle of which coincided approximately with the wheel. There was some improvement on the current wheel, but it was only 10% to 20% efficient. They were mostly used for sawmills. For low heads, they were built low and wide; for higher heads, larger and narrower to make about 120 revolutions or strokes of an upright saw blade per minute.*

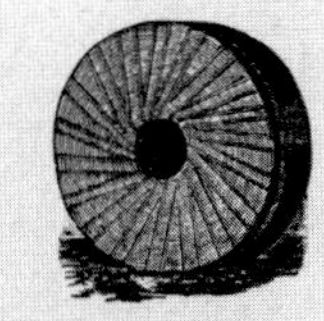

**Price of best selected Old Stock French Burrs.**

**Including Cast Iron Eye, Balance Rine and Patent Balance Boxes.**

| Diameter. | Faced and Backed. | Furrowed. | Diameter. | Faced and Backed. | Furrowed. |
|---|---|---|---|---|---|
| 24 inches. | $45.00 | $60.00 | 42 inches. | $155.00 | $175.00 |
| 30 " | 75.00 | 90.00 | 48 " | 170.00 | 195.00 |
| 36 " | 105.00 | 120.00 | 54 " | 200.00 | 220.00 |

A page from our Improved Success Catalogue published about 1885, showing prices for mill stones. The revolving stone, usually the upper one, is called the "runner" and is much thicker than the lower one, the "bed stone" which is stationary.

*The undershot wheel was an outgrowth of the current and flutter wheels. They averaged about 30% efficiency, and some effort was made in later types to shape the buckets like a trough to hold the water in the buckets longer instead of flat paddles as in the previous wheels. The breast wheels were an improvement on the undershot, the water being applied to the wheel at a higher elevation, usually about the elevation of the shaft. The efficiency varied widely according to the elevation of the application of the water to the wheel, varying from about 35% for a "low breast wheel" to about 65% for a "high breast wheel" in which the water was introduced at an elevation above the shaft centerline. The wooden overshoot wheel was the most efficient of the early wheels, from 65% to 75%. This type was used for the higher heads—15 to 20 feet or more—and the diameter would be a little less than the height of the fall. The wooden overshots were superseded by steel wheels with curved buckets, which gave 80% to*

*90% efficiency. Many of the overshots in York County were built in Hanover by the Fitz Water Wheel Company, which was founded in 1840 and pioneered the steel overshot wheel. The paddle type wheels all had horizontal shafts, but there were also vertical shaft wheels such as "tub wheel" which had an efficiency of 25% to 35% and later a "central discharge wheel," which was equipped with a crude scroll and had an efficiency of from 30% to 40%.*

*Most of the early water wheels of York County must have been of the above types, and there were a great many of them, each mill becoming to some extent the center of life for a small community, and some sites eventually becoming good-sized villages and towns. On a map of York and Adams counties, published in 1821 by D. Small and W. Wagner of York, which may be seen at the Historical Society [present day York County Heritage Trust], the mills of that day are all marked. There were 52 mills of various types on the Codorus and its branches alone, ten on the Little Conewago, ten on the Big Conewago (in York County only), eight on the Bermudian (in York County only), eight on Kreutz Creek and feeders, 15 on Muddy Creek and branches, and 22 others in the county, making a total of about 125.*

*A turbine, as distinguished from a water wheel in the older sense, is a wheel in which motion of the water relative to the bucket is essential to its action. The first scientifically designed turbines in America are usually credited to Francis and Boyden starting about 1850. These men made full-sized drawings of the buckets and studied the path of a particle of water for different velocities under different heads, much as we do today. There was then, and still is, a certain amount of cut-and-try, since to determine the velocity, one must know the discharge, and that is the very thing that is not known until the turbine is built and tested.*

*From about 1860 to 1890, a great many manufacturers, having iron foundries and seeing a chance to sell iron in the shape of a runner for one dollar a pound, entered the turbine field, and a large number of wheels of all types and combinations of types were made. Some of these manufacturers did not know anything about scientific principles of a runner, and many of the types produced gave only 50% to 75% efficiency. It was to correct this condition that the testing flume in Holyoke, Massachusetts, was established in 1879, and even earlier than this, James Emerson made many tests of turbines in Lowell, Massachusetts.*

*Such were the conditions when S. Morgan Smith founded our company in 1877 and entered the water wheel business. His first turbine was the Success, and with its subsequent improvements, the Improved Success No. 1 and No. 2 and the New Success, he obtained an efficiency of over 81% according to Emerson's Tests. In 1889 he obtained the rights to manufacture and sell the McCormick turbine, which was of higher speed, and efficiency of 89%—the best turbine of its day.*

*Some of Mr. Smith's earliest sales were made in York County. His first recorded order for a turbine in York County, according to pages four and five of our Order Book # 1, was on April 22, 1877, from David Bentzel and his son, Edward D. Bentzel, for two 27" waterwheels to run "against the sun"–price $300. Bentzel's Mill was built in 1833 and was on the Little Conewago Creek at the Foustown Road. The wheels were mounted vertically in an open stone flume. One of the 27" wheels was replaced by an 18" No. 2 Improved Success turbine purchased July 24, 1886, Edward D. Bentzel for $95 and probably about the same time the other one was replaced by a Leffel wheel. Edward D. Bentzel lived near the mill, but practiced law in York, having several mill hands operate the mill for him. These mill hands slept in the mill, and took their meals with the Bentzels. About 1916 Chester Werner bought this mill and from 1917 to 1919 Bob Miller of our Dept. B helped him operate it. In 1923 Werner sold out to the late*

*P.A. Elsesser. The mill has not been operated for about the last ten years, but the wheels are still in place.*

*S. Morgan Smith sold not only water wheels, but also other equipment for mills of York County. In a catalog issued about 1890 covering mainly Smith Improved Success Water Wheels, there are tabulations of sizes and prices for gate hoists, rack bars, all kinds of bevel and spur gears and mountings, pulleys, shafting, bearings, roller mills, millstones, corn crushers, elevator cups, smut and separating machines, wheat scouring and polishing machines, flour bolts (sifters), middling purifiers, bran dusters, saw mills and many other smaller items of milling machinery.*

*The mills of York County were not all for grinding grain into meal or flour. There were many saw mills, the first with vertical saws, and later circular saws; tannery mills where water power was used to grind oak bark; oil mills where flaxseed was rolled, ground between a pair of burrs, cooked and pressed to make linseed oil; fulling and carding mills (woolen mills); mills for nail factories, distilleries, and iron furnaces; iron rolling mills; paper mills; and mills for practically every type of manufacturing of the period.*

*Special mention should be made of the York Haven Mills, which were among the largest in the state. After the building of the Conewago Canal around the Conewago Falls in the Susquehanna River in 1795, there was a flouring mill built in 1797 by a Philadelphia Company known as the Conewago Canal Company. Later, in 1810, this company sold out to a Baltimore concern, which built three new mills, one of which had a capacity of 150 barrels per day. ( A barrel of flour is 196 pounds.) The canal and mills became the center of a thriving community and the town of York Haven was laid out on the site in 1814. Keel boats carrying as much as 1,000 bushels of wheat each came down from the north, unloaded at the mills along the canal and were poled back up the river lightly loaded with manufactured goods. After the wheat was milled, the flour was hauled in wagons to Baltimore, and thirty to forty teams with Conestoga wagons would start at one time.*

+++

In February 1961, the Historical Society of York County presented a photographic exhibit that featured Voaden's research on York County mills and waterpower. It was titled "York County Mills, Dams, Streams and Bridges." Never fulfilling his goal of publishing a book on York County mills, Voaden's research files and photographs were given to the Historical Society of York County in 1981. During the 1990s, library and archives volunteers organized the four-drawer filing cabinet of files as well as the hundreds of photographs. Using the name Voaden had chosen when compiling his index, each mill was assigned a number. All of his research on each mill, including the research form containing his notes, was placed in acid-free files. Corresponding photographs were also assigned this number, and they were placed in polyester photo pages. Voaden's index with 14 fields of information was entered into an Excel database file that is now searchable in the York County Heritage Trust library and archives.

Lila Fourhman-Shaull

# CONTENTS

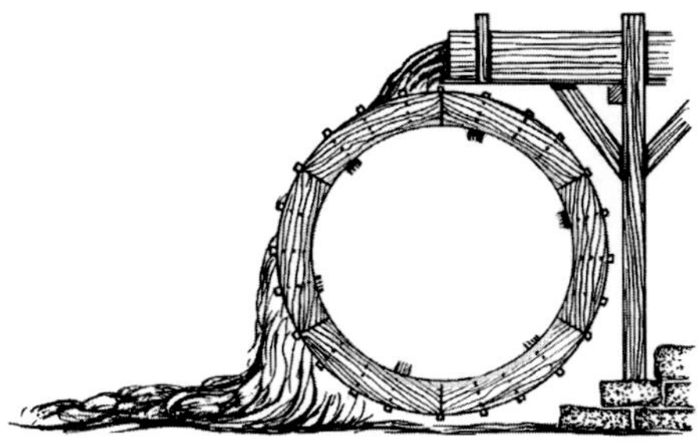

# Beaver Creek

There are two Beaver Creeks in York County. The first is found in the northern part of the county. It begins in Fairview Township and travels southeastward through Newberry Township until it flows into the Conewago Creek. The second Beaver Creek begins in Windsor Township. It divides Lower Windsor and Chanceford townships, where it flows northward joining Fishing Creek, which flows into the Susquehanna River.

**Beaver Creek Mill**

products: feed, corn meal, flour

type mill: grist

property owners: Jonas Feiser, William Feiser, Hantz & Frick, John R. Hershey, Daniel Hoffheins, Carlton Jacobs, George B. Jacobs, Herman Lauber, John Lehn, Jesse Masemer, George E. Myers, Edward E. Sheaffer, John Sinner, John Smith, Samuel Smith, Ludwig Swartz, Abraham Sweigart, Isaac Trimmer, Howard Wallace

Grant Voaden file # 130

**Jacobs' Factory**
products: clover seed, hemp for rope and coarse cloth, lumber, feed and woolen cloth, timber, wheat and rye flour
type mill: grist, saw
property owners: Amos Jacobs, George Jacobs, Isaiah Jacobs
Grant Voaden file # 131

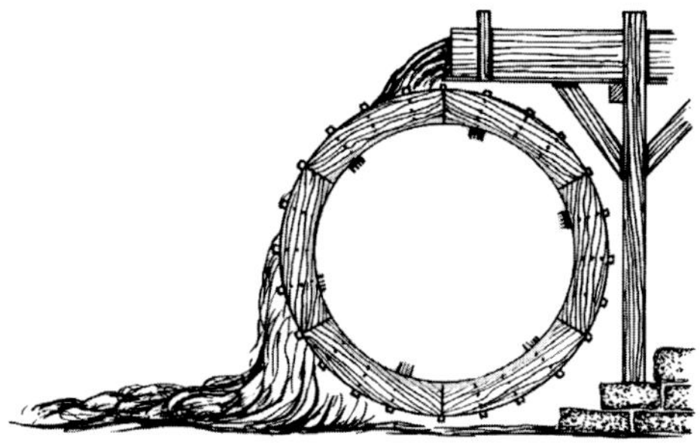

# Bennett Creek

This stream is found on the York County Planning Commission's map but is not identified by name. Bennett Run begins in Fairview Township, north of Lewisberry, and flows southeast across Newberry Township until it enters the Conewago Creek near Camp Cann-Edi-On. Recorded on November 1, 1735 is a warrant for 300 acres along a winding stream to Joseph Bennett, one of the earlier settlers of the Lewisberry area. Authors Mary Elizabeth Bennett Durand and Edward Dana Durand state the following in their 1941 compilation, *The Bennett Family History:* "It was by a descendant of Ellis Lewis that Lewisberry was laid out, and it's from Joseph Bennett that the main stream which winds its devious way through the valley derives its name of 'Bennett's Run.'" The authors further state that this name was in use when the book was published.

**Wilt's Mill**
products: various
type mill: grist
property owners: Levi Fetrow, Aaron Frazer, Aaron Kise, Henry Kise, A. S. Miller, Cyrus Stauffer, Isaac Wilt
Grant Voaden file # 228

**Kauffman's Mill**
products: flour, feed, rough lumber
type mill: grist, saw
property owners: Andrew Cline, Lewis Cline, John Grove, John Harman, Jacob Kauffman, John Kauffman, William Passmore, Herman Updegraff
Grant Voaden file # 230

**Gobrecht Mill**
products: various
type mill: fulling
property owners: Voaden research incomplete
Grant Voaden mill # 234

**Miller's Mill**
products: various
type mill: grist, saw
property owners: Fink, Keister
Grand Voaden file: #264

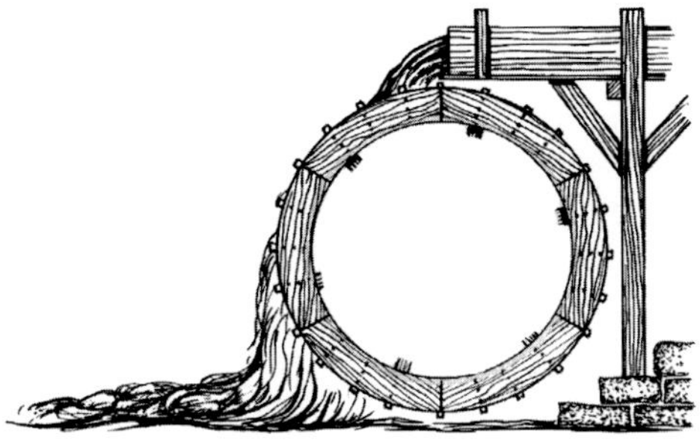

# Bermudian Creek

The Bermudian Creek begins in Washington Township and flows eastward to the Conewago Creek. The origin of this name is unknown. An early reference to this creek is found in the 1940 translation of the Lower Bermudian Church register, which states the following: "the CHURCH BOOK of the Reformed and Evangelical Lutheran Congregations, both true Protestant churches in accordance with the Word of God On the Pramothien (Bermudian) in Hunting (Huntingdon) Township In the Province of Pennsylvania ... ."

**Trimmer's Mill**
products: White Loaf flour
type mill: grist saw, woolen
property owners: Calvin Deardorff, Jacob Eisenhart, Ben Fahnestock, Borius Fahnestock, Dietrich Fahnestock, George Fahnestock, John Groff, Peter Musselman, Smout, John A. Spangler, Martin L. Strayer, Absolom Trimmer, Welker
Grant Voaden file # 120

**Baker's Mill**
products: corn meal, feed, lumber, timbers, woolen goods
type mill: grist, saw
property owners: Calvin Baker, Samuel Baker, Daniel Brubaker, Jacob Brubaker, William Harlacher, Jesse Krall, Samuel Kunkle, Cornelius Leathery, Milton Rodkey, Maurice Seiphert, Lewis Strayer, Thomas Wentz
Grant Voaden file # 122

**Bentz's Mill**
products: various
type mill: clover, grist, saw, woolen
former owners: Carson Bentz, Lewis Bentz, Joshua Deardorff, Joseph Pressel, Valentine Pressel, Ralph Stine, Richard Tully
Grant Voaden file # 124

**Clear Spring Mill**
products: corn meal, feed
type mill: grist, saw
property owners: Edward Dick, George Dick, Scott A. Dick, Adam Ernst, Adam Ernst Jr., John Miller, Joseph Miller
Grant Voaden file # 129

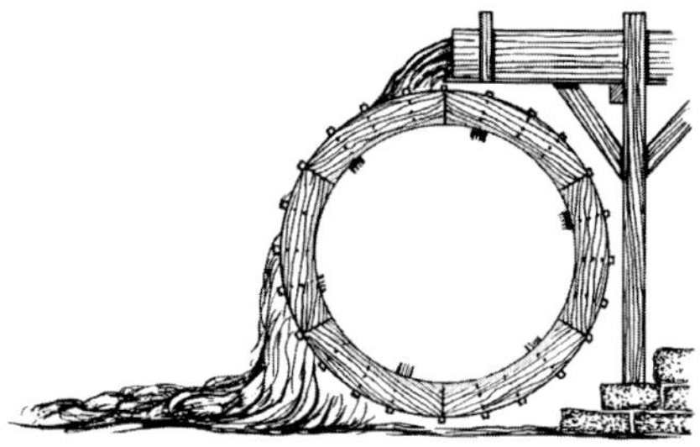

# Cabin Branch Creek

Cabin Branch Creek, shortened to Cabin Creek, begins in Windsor Township and moves eastward through Lower Windsor Township to Long Level, flowing into the Susquehanna River. H. C. Frey speculates that this name may be attributed to the Indians building their "cabins" (log houses) along the Susquehanna River at the mouth of the stream.

**River View Mills**
products: Silver Lace flour
type mill: chop, grist
property owners: John Barr, Christopher Burg, Phillip Burg, Christopher Dosch, Jacob Dritt, John Dritt, Arthur Frey, David Frey, Martin Funk, Andrew Kauffman, Daniel Leber, John Meyer, Christian Rathvon, Christian Witmer
Grant Voaden file # 20

**Margaretta Mill**
products: cider, corn meal, custom feed, Silver Twig flour
type mill: flour, seed
former owners: Jesse Engels, William Himes, Kerns, Benjamin Mittel, Henry Slaymaker, Samuel Slaymaker
Grant Voaden file # 21

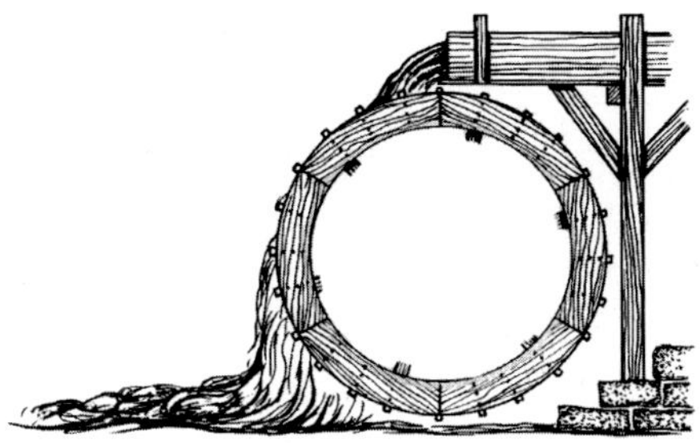

# Codorus Creek

Codorus, possibly of Iroquois origin, means "rapid water." This stream diagonally crosses almost all of York County. It served as the location for the settlement of Yorktown, also known as Little York, York Boro, and present-day York City. The register of York's Christ Lutheran Church contains the following, as translated from the original German text: "Record of children baptized in the Evangelical Lutheran Congregation on the Katores." Other spellings found within records for this stream are Cadores and Cotorus.

This creek, according to the Codorus Creek Watershed Assocation, drains 278 square miles of southwestern and central York County with over 447 milcs of streams. The main branch of the Codorus forms from the junction of two of its extensions, the south branch and the west branch, near the Indian Rock Dam. Continuing northeast through York, it forms the boundary between Springettsbury and East Manchester townships, finally flowing into the Susquehanna River just east of the Codorus Furnace.

Numerous branches flow into the Codorus Creek. There are two "east" branches, one of which joins with the south branch. Eventually, the south branch joins with the west branch. Tyler Run is a small tributary that flows into the Codorus.

Oil Creek begins southeast of Hanover in Penn Township flows north eastward through Heidelberg Township until it joins the Codorus Creek at Menges Mills.

**Codorus Forge and Furnace**
products: pig iron, stove plates
type mill: forge and furnace
property owners: William Bennett, Henry Grubb, Charles Hamilton, Samuel Jago, John Wilkes Kittera, Thomas Niel, Hugh North, James Smith
Grant Voaden file # 44

**Codorus Flint Mill**
products: powdered flint used in pottery manufacturing
type mill: flint
property owners: William Bennett, Curvin Gingerich, Jacob Gingerich, Henry Grubb, Charles Hamilton, Samuel Jago, John Kittera, Thomas Niel, Heber Smith, James Smith, Jacob Wambaugh
Grant Voaden file # 45

**Mundis Mill**
products: Gold Band Feeds, White Lily flour, rabbit pellets, dog food, corn meal
type mill: grist
property owners: Christian Bixler, John Bixler, Joseph Bixler, John Blessing, John Brillinger, Michael Gingerich, Abraham Landis, Andrew Magill, Jacob Mohr, William Mundis, Mathias Rudy, Samuel Rudy, Albert Smyser
Grant Voaden file # 47

**Myers Mill**
products: various
type mill: chop, flour, grist
property owners: Samuel Hershey, Roy Mundis, Edward Myers, Samuel Myers, P. A. & S. Small, York Roller Mills
Grant Voaden file # 48

**Codorus Mill**
products: King of the North, Old Faithful, White Rose, Just Right, Guest, Codorus flour
type mill: grist
property owners: George Small, P. A. & S. Small
Grant Voaden file # 49

**Loucks Codorus Mill**
products: Codorus Brand flour
type mill: grist
property owners: George Albright, Philip Albright, Charles Dempwolf, Samuel Hershey, Conrad Leatherman, George Loucks, Henry Loucks, Zachariah Loucks, Bartholomew Maul, Myers, James Rankin, Small, George Stevenson, Daniel Waggoner
Grant Voaden file # 50

**King's Grist Mill**
products: various
type mill: grist
property owners: Jacob King, Philip Jacob King
Grant Voaden file # 52

**King's Paper Mill**
products: high-grade writing paper
type mill: paper
property owners: Jere Horton, Alfred Jessop, George King, Philip Jacob King, Schmidt & Ault Paper Co., York Water Co.
Grant Voaden file # 53

**Hoke's Mill**
products: feed, flour, lumber
type mill: grist, saw
property owners: Daniel Ebert, Henry Ebert, Martin Ebert, Michael Ebert, Philip Ebert, John Fitzsimmons, Henry Hantz, Daniel Hoke, Martin Hoke, Peter Hoke, David Horn, John Hough, John Menges, Wentworth Vedders
Grant Voaden file # 54

**Rouse Woolen Mill**
products: blankets
type mill: carding
property owners: Rouse family
Grant Voaden file # 150

**Violet Hill Mill**

products: June Bug feed, White Loaf Fancy flour, corn meal, buckwheat flour
type mill: grist
property owners: John Dietz, William Felty, Enos Frey, Daniel Hartman, John Kierman, Peter King, Rufus Kotterman, Jacob Kunkle, Charles Lafean, Joseph Motter, Jacob Newman, Thaddeus Owings, Edward Rouse, Henry Sechrist, Killian Small, Spring Pond Flour Mill
Grant Voaden file # 151

**Moul Mill (found on Oil Creek)**
products: Voaden's information incomplete
type mill: Voaden's information incomplete
property owners: Voaden's information incomplete
Grant Voaden file # 236

**George G. Jacobs Mill (found on Oil Creek)**
products: various
type mill: grist, saw
property owners: Eichelberger, Hartman, George Jacobs, John Jacobs, Roy Jacobs
Grant Voaden file # 237

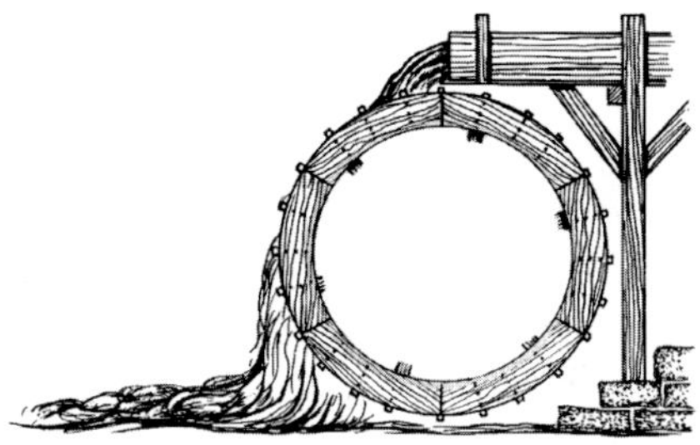

# Codorus Creek, east branch

There are two "east" branches. The first branch forms in Manheim Township, flows eastward to Glenville and northwest until it merges with the "west" branch near Lake Marburg. The second "east" branch begins northeast of Shrewsbury Borough in North Hopewell Township, flows northward through Spring Valley County Park toward Jacobus where it becomes the basis of Lake Redman and Lake Williams. There it becomes one with the south branch.

**Keystone Factory and Grist Mill**
products: flour, feed, woolen goods, lumber
type mill: grist, saw
property owners: Franklin P. Dietz, John Dietz, Enos Frey, Casper Loucks, William Loucks, George Wineholt, Zachariah Wineholt, York Water Co.

Grant Voaden file # 77

**Kreidler's Mill**
products: various
type mill: saw, grist
property owners: Jacob Brillhart, Henry Dollman, Howard Inners, Henry Kreidler, Miller, George Reever, Adam Shearer, Dewald Shearer, Jacob Shearer, John Dewald Shearer, Solomon Shearer, Henry Sterner, Jacob Yost
Grant Voaden file # 78

**Tempest Ness Mill**
products: Codorus flour
type mill: grist, merchant
property owners: Jacob Brillhart, Peter Burgard, Samuel Flinchbaugh, H. R. Godfrey, Jacob Keiser, Tempest Ness, Nicholas Sins, Michael Sintz, Ulrich Sipe, Baltzer Spangler, Jacob Spitler, Michael Valentine, York Water Co., Jacob Yost
Grant Voaden file # 80

**Hopewell Mills**
products: white flour, buckwheat flour
type mill: grist, saw
property owners: George Albright, George Eveler, Sterling Floyd, John Albert Godfrey, Michael Henry, Jacob Jones, Daniel Keeney, George King, John Ruhl, Roy Shaffer, James Sinclair, Samuel Sinclair, William Sinclair
Grant Voaden file # 83

**Hain's Saw and Hominy Mill**
products: hominy
type mill: saw
property owners: Voaden research incomplete
Grant Voaden file # 89

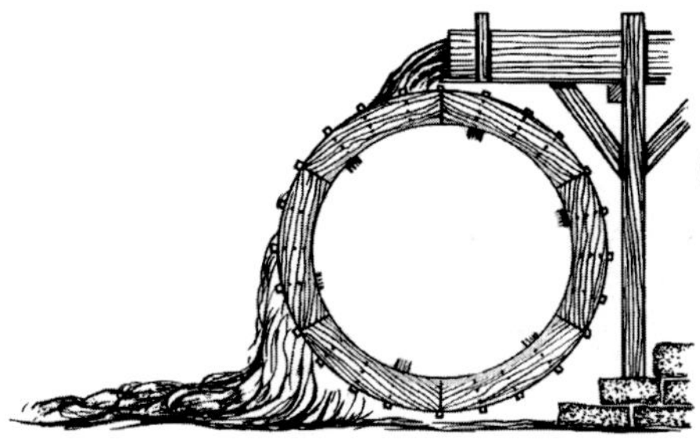

# Codorus Creek, south branch

The south branch begins near New Freedom. It flows through Railroad and Glen Rock and forms the boundary between Codorus and Springfield townships, heading through Seven Valleys until it joins with the west branch.

Trout Run originates in Shrewsbury Borough and flows westward into the south branch of the Codorus Creek, south of Glen Rock.

**Day's Mill**
products: cider, feed, flour
type mill: grist
property owners: John Allison, Nicholas Bupp, Harvey Grimes, Yost Kuhl, George Leber, William Root, John Sprenkle, Levi Sprenkle, Peter Sprenkle, Samuel Weiser, York Water Co., Ira Zeigler
Grant Voaden file # 55

**Gladfelter's Mill**
products: flour, feed, corn meal, lumber, timbers
type mill: grist, saw
property owners: Dietrich Brubaker, Jesse Bubb, Adam Diehl, Daniel Diehl, Jesse Gladfelter, Nathan Gladfelter, Spurgeon Heathcote, Daniel Henry, Washington Henry, John Hummer, George King, Peter King, Peter Menges, George Walter, Daniel Ziegler
Grant Voaden file # 57

**Israel Gladfelter Mill**
products: flax, grain, oil
type mill: fulling, grist, woolen
property owners: Israel Gladfelter, Glen Rock Mill, William Heathcote, Charles Olp, R. Frank Owings, George Seitz, Philip Sheffer, William Sheffer
Grant Voaden file # 59

**Seitzville Flour Mill**
products: flour, feed
type mill: grist
former owners: Henry Caslow, C. C. Chinnault, C. G. Chinnault, Albert Nace, Prowell, Nicholas Seitz, Stephen Seitz, John Slyder
Grant Voaden file # 60

**Larue Mills**
products: flour, feed, lumber until 1905
type mill: grist, saw
property owners: M. A. Bateman, Edward Brodbeck, Christian Henry, George Henry, Jacob Henry, John Kidd, David Martin, Charles Olp, R. Frank Owings, Jacob Schaeffer, Franklin Snyder, Harvey Stine, John Winemiller, John Zeigler
Grant Voaden file # 61

**Radcliffe Rope Mill**
products: flour, feed, rope
type mill: grist, rope (binder twine, sash cord, wash line, cord)
property owners: Centreville Rope Mill, Hanover Cordage Co., Mark Radcliff, George Shaw
Grant Voaden file # 62

**Heathcote Woolen Mill**
products: woolen cloth, yarns, felts for paper industry
type mill: woolen
property owners: James Heathcote, Lewin Heathcote, Spurgeon Heathcote, William Heathcote
Grant Voaden file # 63

**Koller Saw Mill**
products: lumber, timbers
type mill: saw
property owners: John George Ehrman, William Heathcote, Simon Koller, Philip Lau, David Sheffer
Grant Voaden file # 64

**Seitzland Mill**
products: fertilizer
type mill: grist
property owners: E. K. Bollinger, Excelsior Mills, Henry Kohler, Noah Markel
Grant Voaden file # 65

**Rosedale Mill**
products: flour, linseed and flaxseed oil
type mill: grist
property owners: Ulrich Hess, Jesse Klinefelter, Simon Klinefelter, Earl Thoman
Grant Voaden file # 67

**Ruby's Mill**
products: flavin
type mill: grist, bark
property owners: Brown, Helb, G. M. Henry, John Klinefelter, G. W. Ruby, Seiling Furniture Co.
Grant Voaden file # 68

**Bortner Flour Mill, found on Trout Run**
products: feed, flour, grain, meal
type mill: grist
property owners: Samuel Anstine, Adam Bortner, James Bortner, John Bortner, Andrew Diehl, Grover Gantz, Henry King
Grant Voaden file # 71

**Brodbecks Mill**
products: flour
type mill: grist
property owners: Catherine Brodbeck, Theodore Brodbeck, Engles, Hershner, Philip Hildebrand, Joshua Kent, Christof Kolter, Harry Markel, Jacob Sheffer, Jacob Smeach, Clovis J. Stabler, Thoman
Grant Voaden file # 73

**Shaeffer Saw Mill**
products: rough lumber, corn meal, cattle feed, carpenter shop
type mill: grist, saw
property owners: Carl Falkenroth, Harry S. Shaeffer, John Shaeffer
Grant Voaden file # 75

**Smith's Mill**
products: Voaden research incomplete
type mill:: Voaden research incomplete
property owners: Voaden research incomplete
Grant Voaden file # 229

**Rockville Mill**
products: various
type mill: roller
property owners: W. P. Miller, William Swartz
Grant Voaden file # 243

**Honeytown Mill**
products: Voaden research incomplete
type mill: Voaden research incomplete
property owners: Voaden research incomplete
Grant Voaden file # 248

**Leppo Mill**
products: various
type mill: grist, saw
property owners: Jacob Garrett, Milton Garrett, John Leppo, Arnold Sherretz, Frederick Sherretz, Ludwig Sherretz, Samuel Wine
Grant Voaden file # 253

**Cornbower Mill**
products: flour, cider
type mill: saw
property owners: Frances W. Cornbower, H. Eichelberger, Spangler
Grant Voaden file # 263

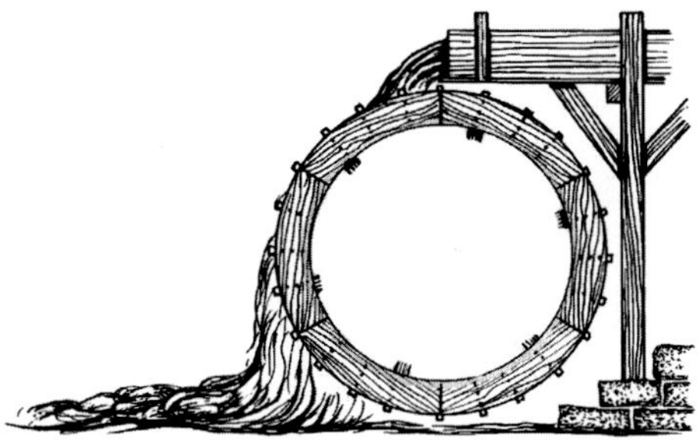

# Codorus Creek, west branch

The west branch of the Codorus Creek begins in West Manheim Township, just north of the Mason-Dixon Line. It becomes the basis of Lake Marburg and forms the boundary between Heidelberg and North Codorus townships. It then flows south of Spring Grove Borough, proceeding northeast to join with the south branch.

**Colonial Valley Mill**
products: various
type mill: grist, saw
property owners: Alvin L. Hershey, Andrew Hershey, Christian Hershey, Peter Menges, Robert T. Menges
Grant Voaden file # 162

**Kraft Grist and Saw Mill**
products: flour
type mill: grist, saw
property owners: Jesse Kraft, William Kraft, John Leib, Michael Leib, Jonas Mummert
Grant Voaden file # 165

**Myer's Mill**
products: flour
type mill: grist, saw
property owners: Fitzpatrick, Kitzmiller, Roy Miller, Milton Myers, William Rheinhart, Jesse Watkins
Grant Voaden file # 166

**Keller Paper Mill**
products: foolscap paper
type mill: distillery, grist, paper, saw
property owners: Jacob Keller, Noah Runkle, Snyder, Richard Staber
Grant Voaden file # 167

**Snyder Flour Mill**
products: Family brand, Blue Ribbon flour, merchant
type mill: grist, merchant
property owners: Amos Belt, Allen Hamme, Jacob Snyder, William Harvey Snyder
Grant Voaden file # 168

**Green Ridge Mill**
products: cider, feed, flour
type mill: grist
property owners: George Bair, E. W. Heindel, Gordon S. Snyder, Titus Snyder
Grant Voaden file # 170

**Codorus Woolen Mill**
products: satin, cloth, cassimeres, jeans, blankets, stocking yarns, flannel, woolen goods
type mill: woolen
property owners: J. F. Bankert, Jacob Keller, John Krebs, Peter Krebs, Runkle, John Sherritz, Ludwig Sherritz, Rufus Snyder
Grant Voaden file # 171

**Glenville Foundry & Machine Co.**
products: tools, cookstoves, ten-plate stoves, plows
type mill: foundry
property owners: Beard & Co., C. M. Brickner, Codorus & Manheim Farmers Association, Glenville Foundry & Machine Co., Jacob Klinefelter, Meyers & Brickner
Grant Voaden file # 172

**Nace's Grist and Saw Mill**
products: cider, feed, flour, linseed oil, lumber, whiskey
type mill: grist, saw
property owners: Daniel Dubs, William Dubs, Harry A. Fry, Erwin Kershner, F. E. Metzger, H. G. Mitzel, William Mitzel, John R. Stine, Adam Warner, J. Park Willey
Grant Voaden file # 174

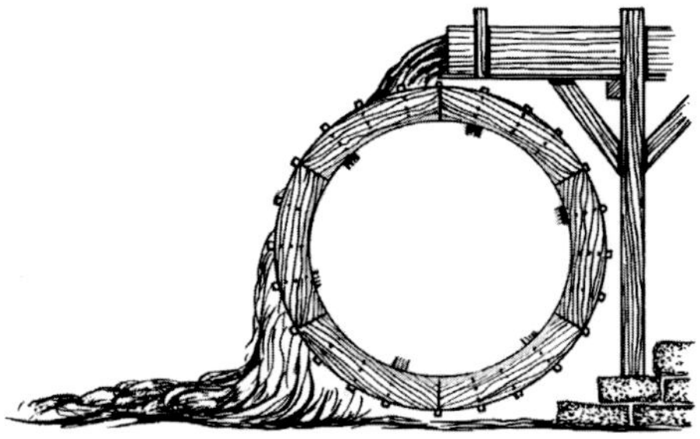

# Conewago Creek

The name of this creek is attributed to the Lenape Indians and means "at the rapids." Stephen G. Boyd in his volume, *Indian Local Names,* defines "gune uage" as meaning a "long strip" or "long reach" and first used for an area near Middletown in Dauphin County.

The Conewago Creek watershed has a total area of 515 square miles, including areas in Maryland and Pennsylvania. This creek and its tributaries drain over 50% of Adams County and nearly 27% of York County. Originating in Franklin Township, Adams County, it weaves around the borough of East Berlin and flows northeast through York County, making its way to the Susquehanna River at York Haven. It forms the boundaries of several York County townships: Washington, Paradise, Dover, Warrington, Conewago, Newberry, and East Manchester.

**Conewago Roller Mill**
products: animal feed, chop, flour, lumber
type mill: roller
former owners: Norman Bair, Henry S. Bear, Jacob Bear, Andrew Cline, James Cline, Samuel Cookson, Henry Demuth, Reuben Drawbaugh, Gregor, Henry Hoff, Harry Hykes, Herbert Kimmel, Abraham Landis, Abraham Meyer, George Shetter, Martin Shetter, Henry Stoehr, William Wertz, John Wilson
Grant Voaden file # 100

**Kunkel's Flour Mill**
products: lumber, chop, feed, flour
type mill: grist, saw
property owners: Belshur Kunkel, Elijah Kunkel, Gottlieb Kunkel, Jacob Kunkel, Peter Kunkel
Grant Voaden file # 101

**Detter's Mill**
products: grist, corn meal
type mill: grist
property owners: Harry Detter, Joseph Reuben Detter, George Emig, Martin Emig, Jacob Frick, Samuel Harlacker, George Sheffer, Charles Spahr, Adam Speck, Amos Swartz, Tempest Tucker
Grant Voaden file # 102

**Bigmount Mill**
products: flour, feed
type mill: grist
property owners: Sarah Diehl, Susan Diehl, George Emig, William Kling, Jacob B. Kunkel, John Ort, C. Elmer Smith, S. Morgan Smith, J. G. Spangler
Grant Voaden file # 103

**Washington Mill**
products: buckwheat flour, cider, corn meal, feed, White Loaf flour
type mill: grist, saw
property owners: Emanuel Butt, Christian Closs, Conrad Eisenhart, Elmer B. Eisenhart, Franklin Eisenhart, Peter Eisenhart, Isaac Hershaw, Isaac Latshaw, Joseph Latshaw, Abraham Myley, Peter Ohler, William Ohler, Jacob Welsh
Grant Voaden file # 104

**Kunkle Grist Mill**
products: various
type mill: grist, saw
property owners: Henry H. Kockenhour, Peter S. Kunkle, Samuel Quick, Ivan Wallick
Grant Voaden file # 109

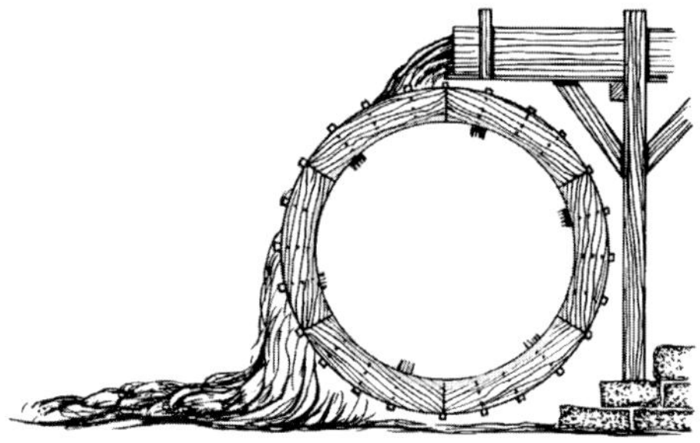

# Deer Creek

Deer Creek originates in Shrewsbury Borough, York County, and flows in a southeasterly direction. It forms the boundary between Shrewsbury and Hopewell townships and crosses the Mason-Dixon Line into Baltimore County, Maryland, continuing to the Susquehanna River.

**Keeney Flour Mill**
products: feed, flour, grain
type mill: saw
property owners: Deer Creek Roller Mills, Jacob Keeney
Grant Voaden file # 223

**Ebaugh Saw and Grist Mill**
products: feed, flour
type mill: grist, saw
former owners: Deer Creek Water & Waterpower, Adam Ebaugh, Clay Ebaugh, John Ebaugh, John Henry Ebaugh, Milton Charles Ebaugh, Dr. Miller, Stewartstown Water Authority
Grant Voaden file # 224

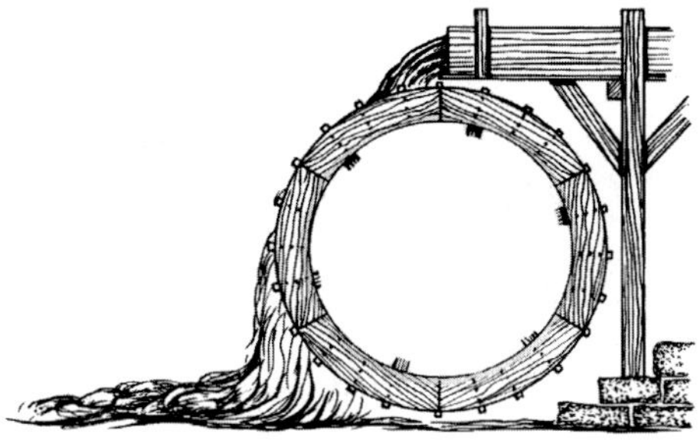

# Fishing Creek, north

There are two Fishing Creeks in York County, and both flow to the Susquehanna River. One is in the northern part of York County and flows from Fairview Township south through northeastern Newberry Township to the Susquehanna River.

**Wentz Mill**
products: various
type: grist, saw
property owners: Gerry Bamberger, James Detweiler, Glenn Ebersole, John Fetrow, Leo Garvey, Christian Hess, Richard Leggett, William Nelson, James Patterson, P. B. Smith, Joseph Tate, Eliza Wentz, Reuben Wentz
Grant Voaden file # 91

**Yocum Woolen Mill**
products: various
type mill: fulling, woolen
property owners: Jacob Arnold, Jesse Berkheimer, Daniel Ginder, William Nailer, Elijah Yocum
Grant Voaden file # 94

**Red Mill Inn (also known as Jolly Coppersmith)**
products: flour
type mill: grist
property owners: Jacob Bare, Copper Kettle, Joseph Prowell
Grant Voaden file # 90

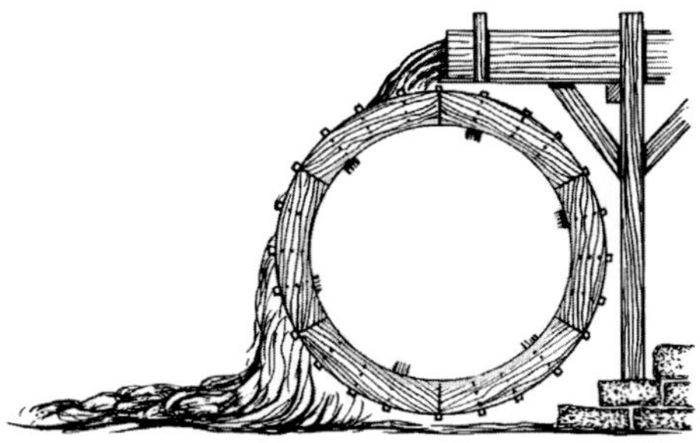

# Fishing Creek, south

The southern Fishing Creek has its origins just southeast of Windsor Borough. From there, it flows eastward through Lower Windsor Township and into the Susquehanna River, south of Long Level.

**Fishing Creek Roller Mill**
products: White Rose flour
type mill: grist
property owners: Elmer Beck, Bentz, Ervin Kohler, Paul Leiphart, Wiley Markel, Smeltzer, Cornelius Snyder, William Wise
Grant Voaden file # 26

**Bahn's Mill**
products: White Lily flour, dog food, special chicken feed
type mill: grist
property owners: Daniel K. Anstine, Simon Anstine, Simon B. Anstine, Simon H. Anstine, Charles Bahn, Emanuel Bahn
Grant Voaden file # 27

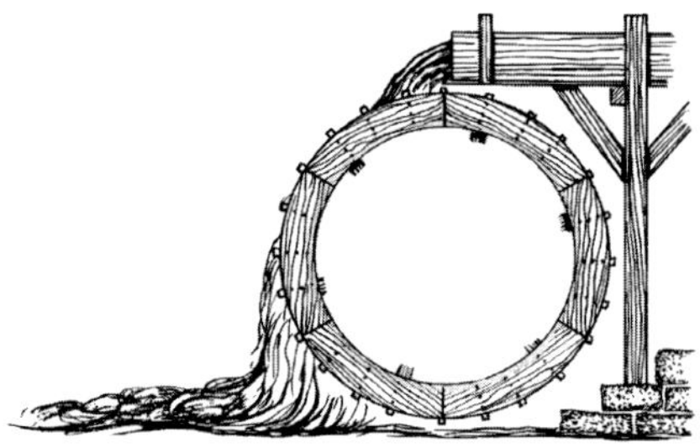

# Kreutz Creek

Kreutz Creek begins in Windsor Township flowing north and east through Hellam Township to the Susquehanna River. John Grist settled on unwarranted land just north of this stream where it empties into the Susquehanna River in 1719.

"Most of the early German immigrants settled in the neighborhood of the Kreutz Creek," according to Carter and Glossbrenner's *History of York County*.

Grist's Creek is first mentioned in a 1735 Blunston license, and two other variations — Grises Creek and Grices Creek — are found in the Maryland Archives.

**Strickler (Henry) Mill**
products: various
type mill: chop, grist, saw
property owners: William Flora, James Houston, Robert Houston, Susanna Wright Houston, Andrew Kauffman, Christian Kauffman, Henry McElroy, Pennsylvania Joint Lumber, Anson Phelps, Henry Reigart, Jacob Ruth, Christian Stoner, Henry Strickler, Horace Washburn, John Wright, Wrightsville Power and Light
Grant Voaden file # 1

**Dietz/Bair Mill**
products: buckwheat flour, clover seed, corn meal, flour
type: grist, saw
property owners:Woodrow Bair, John Davis, Calvin Dietz, George Dietz, Thomas Doyle, William Jones, Henry Strickler, Jacob Strickler, John Strickler, Ulrich Strickler, William Strickler
Grant Voaden file # 2

**Willow Grove (also known as Mackley's Mill)**
products: various
type mill: grist, still
property owners: George Ebert, Andrew Ernst, Acquilla Fauth, John Hershey, John Hertzler, Abraham Hiestand, John Hiestand, Jacob Huber, Zachariah Loucks, Philip Mackley, George Mann, John Mann, Philip Mate, John Miller, Henry Shultz, Martin Shultz, Elwood Sprenkle, Christian Stoner, Weltzhoffer
Grant Voaden file # 5

**Stoner Distillery**
products: whiskey
type mill: distillery
property owners: John Stoner
Grant Voaden file # 6

**Locust Grove Mill**
products: flour, Locust Grove Best Rose, Pure Winter Wheat flour
type mill: saw, grist, distillery
property owners: Walter Beard, Joseph Hursh, Joe Kindig, George Libhart, Henry Libhart, John Mate, Philip Mate, Christian Ratfohn, Michael Rudy, Shue, John Stoner, Daniel Stouffer, Jacob Stouffer, Robert Swengl, Jacob Welshoffer
Grant Voaden file # 7

**Graybill Clover Mill**
products: clover seed
type mill: clover
property owners: Elmer Dietz, Henry Dietz, John Dietz, David Graybill, Andrew Hinkle
Grant Voaden file # 8

**Gehley's Woolen Factory**
products: stocking yarn, flannels, blanket, clothes
type mill: fulling, woolen
property owners: Daniel Gehley, John Gehley, Eli Hersh, John Mathews, Robert Myers
Grant Voaden file # 12

**Miller's Mill**
products: Lily White flour, feed, lumber, corn meal
type mill: grist, saw
property owners: Andrew Miller, Michael Miller, Henry Schrantz, Christian Stoner, Samuel Stoner, Henry Tyson, Wineka
Grant Voaden file # 13

**Sebastian Stevens Mill**
products: flour, feed
type mill: grist
property owners: Eli Kindig, John J. Steinfelt, Sebastian Stevens
Grant Voaden file # 15

**Crumbling Grist Mill & Store**
products: phosphates, grain, feed, flour, fertilizer, cider press
type mill: grist
property owners: Benjamin Franklin Crumbling, George Crumbling, John Kauffman, Melvin Sechrist
Grant Voaden file # 17

**Throne (Hieth) Grist Mill**
products: corn meal, cider, feed
type mill: grist
property owners: David Brubaker, Buchart, Susan Glatz, Isaac Groff, Christian Hammacher, Abraham Hiestand, William Hiestand, Knobb, Dr. Francis Koch, Christian Miller, Christian Oberholtzer, John Shultz, Hieth Trone
Grant Voaden file # 18

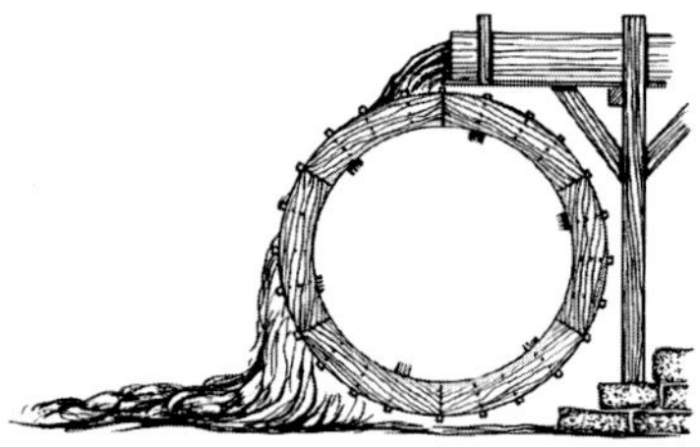

# Little Conewago Creek

The little Conewago Creek begins in Jackson Township near the York Airport and flows northeast, eventually forming the boundary between Dover and West Manchester townships as well as the boundary between Conewago, Manchester and East Manchester townships. It joins with its counterpart, the Conewago Creek, north of Manchester Borough.

**Kline's Mill**
products: flour, feed, lumber
type mill: grist, saw
former owners: Peter Altland, Blessing, Brenneman, Elmer Fetrow, Fink, Abraham France, William Kline, Rice, Franklin Shaffer, Starr, Swartz
Grant Voaden file # 106

**Conley's Mill**
products: feed, flour, lumber
type mill: grist, saw
property owners: Philip Baker, John Becker, William Conley, John Conolly, William Ewing, John Fink, James Love, Martin Mansberger, Miller, Leonard Swartz
Grant Voaden file # 107

**Old Mill Inn**
products: Bear's Best corn meal, Bear's White Lily flour, feed, pretzel flour
type mill: grist
property owners: George Fink, John Frye, George Gross, Zacharias Gross, Philip Hoffman, George Jordan, Henry Kochenour, Jacob Kunkel, Samuel Lichtenberger, Daniel McLaughtery, John Meisenhelter, George Smyser, William Spangler, John Updegraff
Grant Voaden file # 108

**Bentzel's Mill**
products: flour, lumber, cigar boxes
type mill: grist, saw
property owners: David Bentzel, Edward Bentzel, Feliz Bentzel, Binder, Paul Boring, Emory Eisenhart, Peter Elesser, George Ilgenfritz, Joseph Ilgenfritz, Martin Ilgenfritz, Joseph March, Martin, Meisenhelter, Chester Werner
Grant Voaden file # 110

**Anderson's Mill**
products: feed, flour
type mill: grist
property owners: Romanus Anderson, Charles Axe, Andrew Hake, Levi Hake, Dr. Harry King, Walter Young
Grant Voaden file # 112

**Shettle's Mill**
products: corn meal, feed
type mill: grist
property owners: H. D. Airesman, Amos Bentzel, Bentzel, Abraham Frantz, Jesse Heilman, Henry Kochenour, Arthur Kohler, George Meisenhelter, George Shettle, George Shram, John Stough, Jacob Swartz
Grant Voaden file # 113

**Neiman's Mill**
products: feed, Golden Grain flour
type mill: grist
property owners: Michael Beltzhoover, W. K. Coxen, Robert Emerton, William Emig, Daniel Gross, Samuel Gross, Holtzapple, Karl Katz, L. B. McCullough, George Neiman, John Neiman, Schmucker, Spangler, Walters, Martin Weigle
Grant Voaden # 114

**Emig's Mill**
products: feed, flour
type mill: chop, grist
property owners: Gabriel Emig, Irwin Emig, Jackson Emig, Jacob Emig, Reuben Emig, Jacob Gross, Kauffman, Lockman, Ludwig Spiess, Peter Spiess, Harman Updegraffe, Andrew Ziegler
Grant Voaden file # 115

**Biesecker's Mill**

products: White Lily & White Swan flour, Yorko Dog Food, Golden Corn meal
type mill: grist, saw, distillery
property owners: John Biesecker, Elmer Eisenhart, Samuel Gross, William Julius, George Myer, Daniel Reicher, Andrew Schwartz, Joseph Smith, Peter Smyser, John Stouffer, Michael Stover, Philip Weis
Grant Voaden file # 116

**Deardorff's Grist Mill**
products: feed
type mill: grist
property owners: Jesse Appleman, Thomas Appleman, Martin Brillhart, David Deardorff, George Falkenstein, August Felty, William Felty, Luther Galletin, Samuel Gross, John Jacobs, Susquehanna Pipe Line Co.
Grant Voaden file # 118

**Spangler's Mill**
products: feed
type mill: grist
property owners: Daniel Spangler, George Spangler, William Spangler, Thomasville Lime & Stone Co.
Grant Voaden # 119

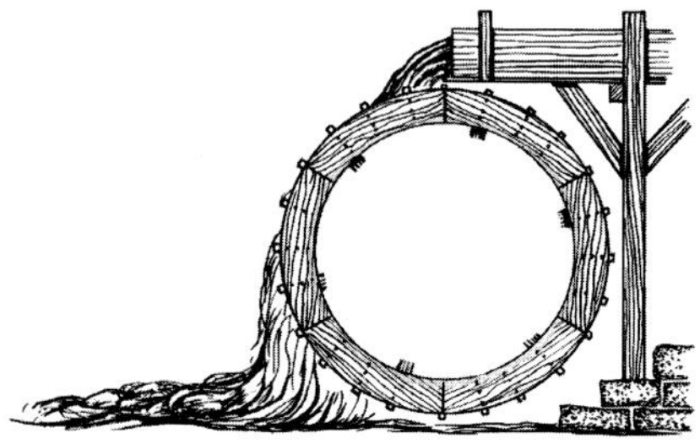

# Mill Creek

This creek has its origins outside Red Lion. Also known as the Little Codorus, it flows through Dallastown, northwest through Ore Valley, along Route 83, and around York City until it joins with the main branch of the Codorus Creek.

**Diehl's Mill**
products: meal, feed, flour
type mill: grist
property owners: Adam Diehl, Daniel Diehl, Jacob Diehl, Peter Diehl, Enos Hively, William Hively, Adam Leitner, Henry Louch, Mathias Sitler, Chester Werner
Grant Voaden file # 136

**Horn's Grist Mill**
products: corn meal
type mill: grist
property owners: East York Mill, John Flory, George Heidlebaugh
Grant Voaden file # 138

**Elmwood Roller Mill**
products: Elmwood White Loaf Flour
type mill: grist
property owners: C. B. Landes (leased), John Small
Grant Voaden file # 139

**Plank Road Mill**

products: various

type mill: custom feed, saw

property owners: Michael Bare, John Bergerhouse, John Davis, John Christopher Dosch, John Harbaugh, Conrad Holtzbaum, Daniel Jacoby, Israel Kauffman, John Kauffman, John Landis, Daniel Lefever, George Lefever, John Lefever, Penn Fibreboard, John Rorer, David Sherich, Jacob Small, John Small, Lydia Small, Alfred Strickler, Ulrich Wisler

Grant Voaden file # 140

**Spring Garden Mill**
products: various
type mill: flour
property owners: P. A. & S. Small
Grant Voaden file # 141

**Bowser's Cider Mill**
products: various
type mill: cider, grist
property owners: Martin Austin, Israel Bowser, Clair Meyers, Rider, Ernest Yost
Grant Voaden file # 143

**Strickler's Mill**
products: iron pans and ladles, forge for manufacturing of iron implements
type mill: forge, grist, saw
property owners: George Musser, Harry Strickler
Grant Voaden file # 145

**Ore Valley Station Mill**
products: flour, feed, lumber
type mill: grist, saw
property owners: Henry Caslow, Peter Diehl, William Peters, James Raby, Edward Reigart, Samuel Small, Albert Stiles, Henry Strickler, Jacob Taylor
Grant Voaden file # 146

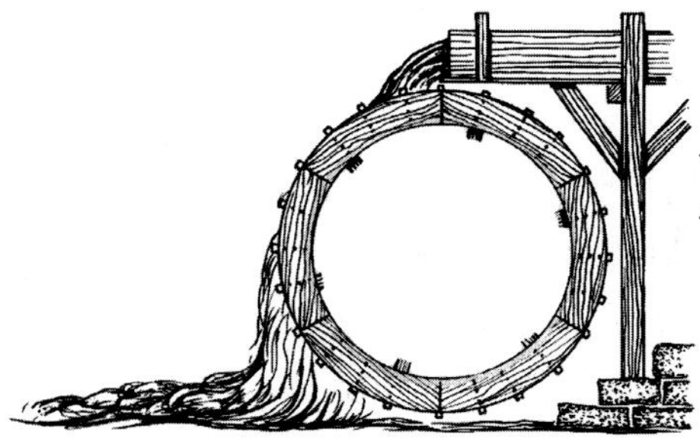

# Muddy Creek, north branch

Muddy Creek originates in Windsor Township. Boyd states that this stream had been called by the Indians achsee-pagkoh, i.e. “muddy water.” The north branch of Muddy Creek forms the boundary between Windsor and North Hopewell townships as well as East Hopewell and Chanceford townships.

**Bryansville Grist Mill**
products: feed, flour, grain
type mill: grist
property owners: J. F. Gibson
Grant Voaden file # 38

**Gartridge Power Plant**
products: electricity
type mill: electric power
property owners: Delta Electric Power Co.
Grant Voaden file # 183

**Castle Fin Hydroelectric Plant**
products: electricity
type mill: electric power
property owners: Delta Water Power Co. (other names: Delta Electric Power Co., Pennsylvania Water & Power Co., Southern Pennsylvania Power Co., Susquehanna Power Co.)
Grant Voaden file # 184

**Muddy Creek Forks Roller Mill**
products: corn meal, buckwheat, feed, flour
type mill: grist
property owners: James Cooper, William Douglas, Andrew Duncan, Alexander Grove, Joseph Hamilton, James Keiser, Alexander McCurdy, James Maffet, Henry Manifold, Joseph Manifold, John Reed, Dr. John Rouse, Peter Rouse, Andrew Turk, Alexander Turner
Grant Voaden file # 188

**High Rock Mill**
products: Bungalow Family flour, feed, merchant
type mill: grist
property owners: Wallace Dunlap, Oscar Fox, William Shultz
Grant Voaden file # 189

**Felton Station Mill**
products: various
type mill: grist
property owners: Crumbling, Eastern Milling Co., Charles Grove, Horace Hake, George Kohler, H. H. Minnich, Payne, Henry Sechrist
Grant Voaden file # 191

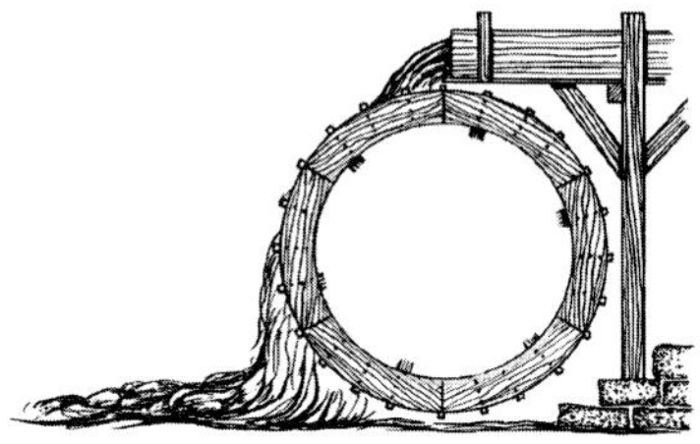

# Muddy Creek — Rambo Run

Rambo Run is found in East Hopewell Township. It begins at the southwestern end of the borough of Cross Roads, flows through the southern end of the borough and ends when it meets with the north branch of Muddy Creek. It received its name from Aaron Rambo, who owned property along this stream in the latter part of the 19th century.

**Cross Grist Mill**
products: feed, corn meal (white and yellow)
type mill: grist
property owners: Harry Cross, J. Nelson Cross, Alexander Wallace, Deborah Wallace, James Wallace, Mary Angeline Wallace, William Wallace
Grant Voaden file # 199

**Hyson Mill**
products: chop, flour
type mill: grist
property owners: Archie Hyson, Foster Keesey
Grant Voaden file # 200

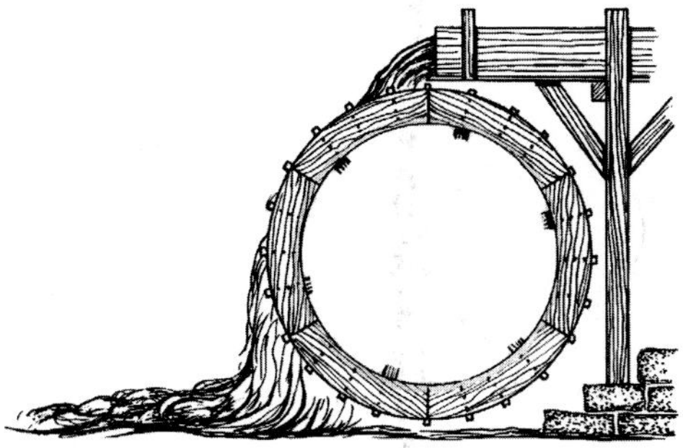

# Muddy Creek, south branch

The south branch of Muddy Creek begins just west of New Park and travels north. It forms the boundaries between Hopewell and Fawn townships and eventually meets with the north branch at Muddy Creek Forks. From there, it flows southeastward to the Susquehanna River.

**McKinley Mill**
products: feed, buckwheat flour, cider
type mill: grist
property owners: Harlan Clayton, William Clayton, Alvey Hess, James Kennedy, George McKinley, J. Harry McKinley, John Reynolds, George Walters
Grant Voaden file # 209

**Gemmill's Mill**
products: flour, feed, cider
type mill: grist
property owners: Lillie Atkins, Barnett, Burk, Gladys Burkins, James Nelson Gemmill, J.Y. Bryan Gemmill, James Hill, George Pitts, Sylvester Pomraning, John Snyder
Grant Voaden file # 210

**Woodbine Station Mill**
products: flour, feed, corn meal
type mill: grist
property owners: James Cooper, William Cooper, John Donnell, David Hoffmaster, George W. Ilgenfritz, William Ilgenfritz, Arthur King, Kenneth Pitzer, Cora Ruff (Mrs. Joe Ady), H. Weldon Ruff, Lula Swift, Henry Wackerman, James Wineholt
Grant Voaden file # 211

**Garvine Mill**
products: flour, feed, corn meal
type mill: grist
property owners: Allen Anderson, William Anderson, E. A. Blaine, Robert Blaine, William Blaine, W. J. Garvine, Ralph Pomraning, Robert Traverse
Grant Voaden file # 213

**Miller's Mill**
products: flour, cider, lumber
type mill: grist
property owners: Maude Ayres, Arthur Black, Henry Christian, Francis Grove, Jacob Grove, Alexander Liggit, Stephen McKinley, Joseph Manifold, Alexander Martin, Samuel Martin, W. E. Miller, Clarence Rhoda, Isaac Terry, W. W. Tompkins, James Wilson, Earl Zirkle
Grant Voaden file # 214

**Grove Mill**
products: Lily Valley flour
type mill: grist
property owners: Edmond Cummings, Ken Durham, Jacob Grove, James Emory Grove, James Frank Grove
Grant Voaden file # 216

**Brooks Mill**
products: various
type mill: grist, saw
property owners: Samuel Brooks, John Garrison, Jacob Grove, Joseph Jordan, William G. Montgomery
Grant Voaden file # 218

**Liggit Mill**
products: various
type mill: grist
property owners: Tiddlings Grove, C. Wilson Liggitt, Samuel Liggit, Joseph Mitzel
Grant Voaden file # 219

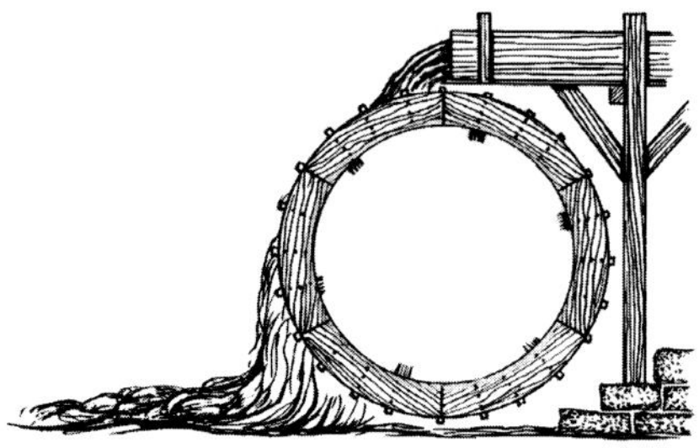

# Otter Creek

Otter Creek, also known as Rapid Creek, originates in the western section of Chanceford Township. It flows east and southeast, entering Lower Chanceford Township, and reaches the Susquehanna River just north of York Furnace.

**Centennial Mills**
products: Buckwheat flour, corn meal, White Lily flour, stock feed
type mill: grist
property owners: Joseph Atkinson, Robert Bacon, Alan Davis, Ervin Kohler, George Kohler, Maynard Lighty, Reed, William Strickler, Harvey Wineholt
Grant Voaden file # 28

**Otter Creek Mills**
products: lumber, corn meal, buckwheat flour
type mill: grist, saw
property owners: Henry Baldwin, Samuel Daughtery, Ivan Keeports, Robert J. Keeports, Colonel George Murphy, James Scott, Leonides Scott
Grant Voaden file # 29

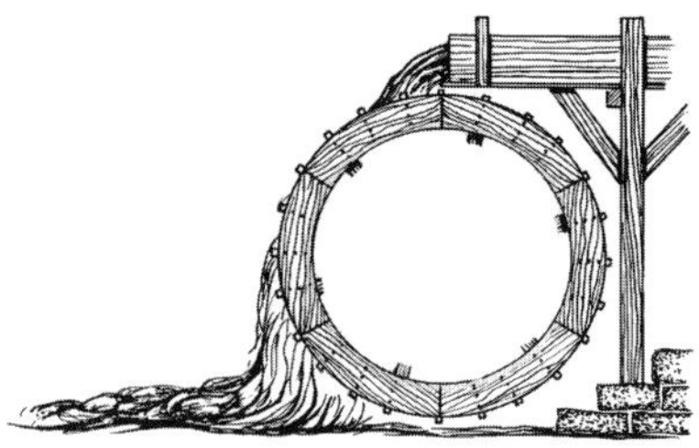

# Susquehanna River

The Susquehanna River is the sixteenth-longest river in the United States. It originates near Cooperstown, New York, crosses Pennsylvania, and flows into the Chesapeake Bay at Havre de Grace in Maryland.

Some state the name Susquehanna is derived from the Algonquin word Susquehanough, the meaning of which varies from "people at the falls" and "people of the Muddy River," to "people of a well-watered land." In his book *Indians in Pennsylvania,* Paul Wallace notes that the name "Susquehannock" was not the name the tribe called itself. Author Jack Brubaker summarized it best in *Down the Suquehanna River to the Chesapeak*e, when he stated there are numerous variations of the Algonquin word "Susquehanna" with much discussion on its definition.

The river forms the eastern boundary of York County and separates it from its parent county, Lancaster.

**Accomac Mill**
products: Voaden research incomplete
type mill: Voaden research incomplete
property owners: Voaden research incomplete
Grant Voaden file # 262

# Appendix

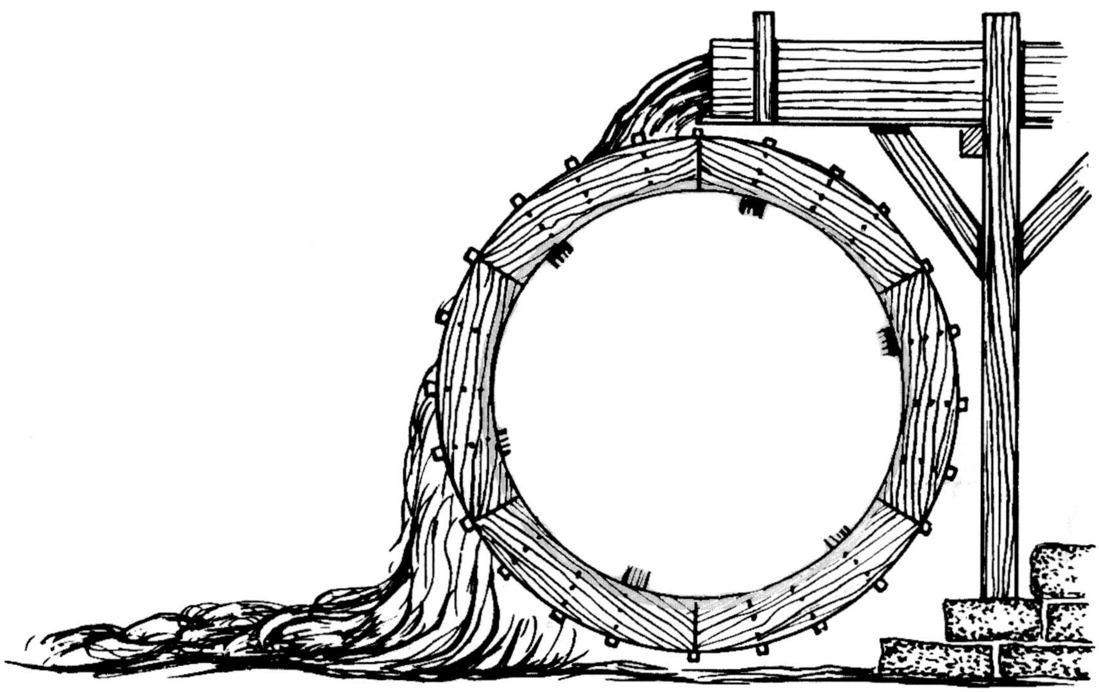

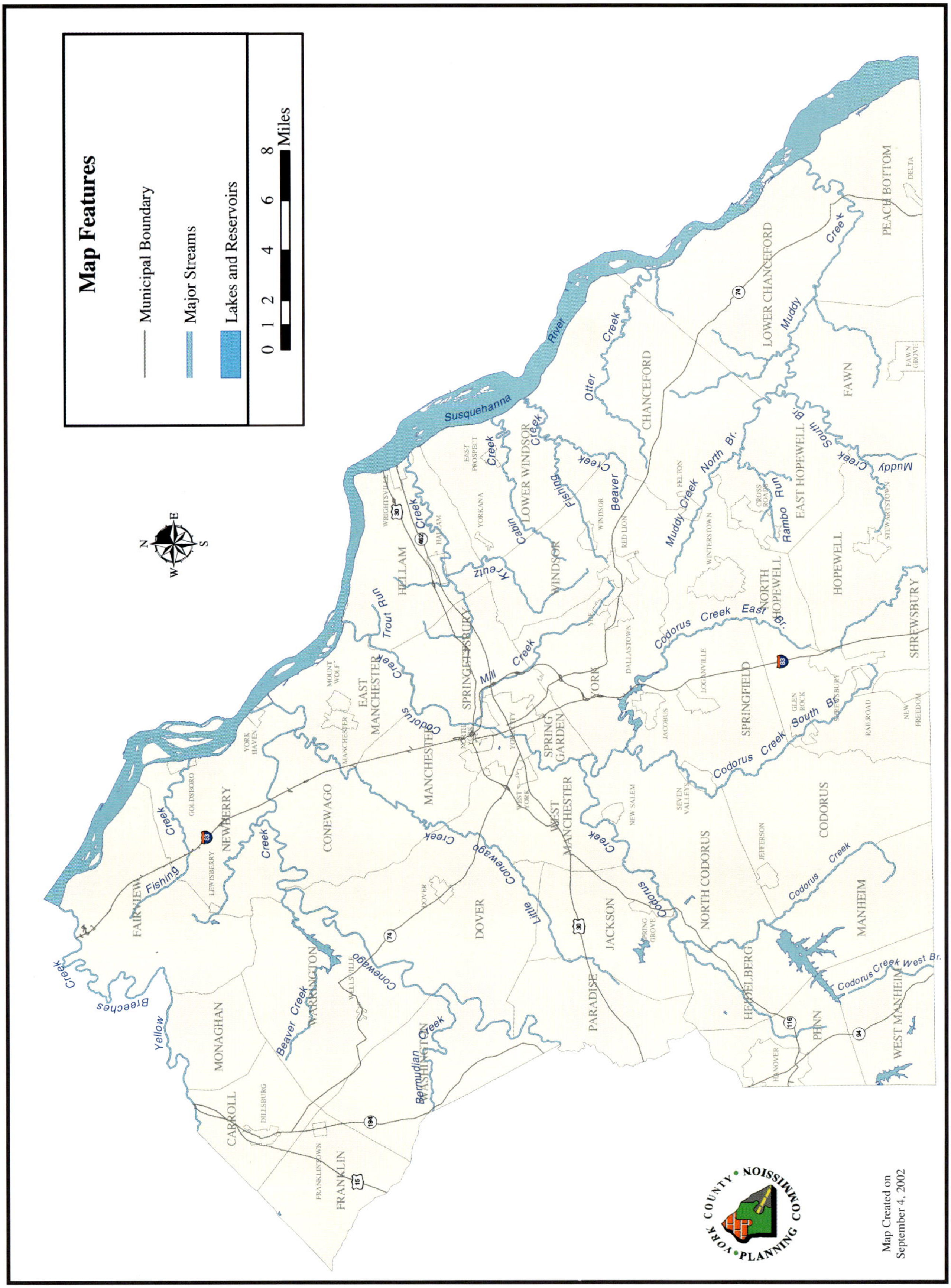


Major Streams in York County, Pennsylvania

As far as we know this is the only copy of this map in existence. The original is the property of the Historical Society of York County, and we thank them for allowing us to reproduce part of it. The original is in color and the mills are marked with small sketches of water wheels. We used stars to make them stand out. Another interesting point is the number of churches. They are all marked with a little sketch of a church as you will see if you study the map.

Map of York and Adams counties published by D. Small and W. Wagner — 1821 found in pamphlet
*Art * Ink & Type*

# Creeks of York County

Whether labeled a creek, a run, a river, or a branch, below is an alphabetical list of the streams presently found in York County. The bold entries are referenced within Voaden's files.

Alum Rock Run
Anderson Run
**Bald Eagle Creek**
**Barshinger Creek**
Bear Branch
**Beaver Creek**
**Bennett Run**
**Bermudian Creek**
Big Spring Run
Boyds Run
Bull Run
**Bunch Creek**
Cabin Creek (**Cabin Branch Creek)**
Canadochly Creek
**Canoe Run**
Carter Creek
**Centerville Creek**
Cherry Run
**Clear Spring Run**
**Codorus Creek**
**Conewago Creek**
Cook Creek
Counselman Run
Cuffs Run
Davidsburg Run
**Dee Run**
**Deep Run**
**Deer Run**
Delta Run
Doe Run
**Dogwood Run**
Dugan Run
Duncan Run
**East Branch Codorus Creek**
East Branch Toms Run
**Ebaugh Creek**
**Emma's Creek**
Falling Branch
**Fishel Creek**
Fishers Run
**Fishing Creek**
Foust Creek
**Fox Run**
**Furnace Creek**
Furnace Run
Gitts Run
**Glenville Branch of Codorus Creek**
**Gray Rock Mill**
Green Branch
Gunpowder Falls
Hartman Run
Honey Run
**Indian Run**
**Inners Creek**
Island Branch
**Jefferson Creek**
Klines Run
**Krebs Valley Run**
**Kreutz Creek**
Laurel Run
Leibs Creek
**Little Codorus Creek**
Little Conewago Creek
Little Deer Creek
**Long Arm Creek**
**Long Run**
Mahala Run
**Manhattan Creek**
Marsh Run
Michael Run

Mill Branch
**Mill Creek**
Millers Run
Mud Run
**Muddy Creek**
**Musser Run**
Neill Run
**North Branch Beaver Creek**
**North Branch Bermudian Creek**
North Branch Muddy Creek
Oakland Run
**Oak Valley Run**
**Oil Creek**
Old Mill Creek
**Orson Run**
**Otter Creek**
**Paradise Creek**
Paradise Run
**Pierceville Run**
Pineville Run
**Pine Run**
Pippins Run
Porters Creek
Railroad Run
**Rambo Run**
Red Run
Robinson Run
**Rock Run**
**Sawmill Run**
Scott Creek
**Scott's Run**
**Seaks Run**
**South Branch Codorus Creek**
**South Branch Conewago Creek**
**South Branch Muddy Creek**
South Fork Otter Creek
**Sowego Creek**
**Stony (Stoney) Run**
Stoverstown Branch
**Susquehanna River**
**Tome's Creek**
**Toms Run**
**Trout Run**
**Tylers Run**
Wallace Run
**West Branch Codorus Creek**
West Branch Toms Run
Wildcat Run
Willis Run
Wilson Run
Wolf Run
Yellow Breeches Creek

# GLOSSARY

APPLE JACK: A local name for fermented apple juice (aka Stone Fence)

ATTRITION MILL: A mill that grinds grain by the rubbing of two corrugated disks traveling in opposite directions. A transition between stone buhrs and hammer mill.

BEDSTONE: The lower stone, which is stationary, of a pair of millstones.

BRAN: The outer layer (skin) of wheat, rye, or oats, separated by grinding.

BREAK: Any of the roller mill processes by which the grain is made into flour. The first break begins the process, followed by the second and as many as the mill has.

BREAST WHEEL: The water reaches the buckets at about the middle of the wheel. The wheel is turned by the impact of the water and its weight.

BURRS: Millstones.

BUHRSTONE MILL: An old gristmill using burrs or millstones for grinding.

CAPSTONE: Another name for the runner stone.

CARDING MILL: Machines used to clean, separate, and process wool.

CHOPPING: The grinding of a mixture of grains.

CONDITIONING: The wetting of the grain to prepare it for grinding by toughening the bran and mellowing the floury part of the grain. OR Tempering.

CORN CUTTER: A machine in the mill to cut grains of corn for poultry feed.

DAMSEL: A device that agitates the hopper, as the upper millstone revolves, so that the grain is shaken down into the stone for grinding.

DRESS: To true the surface of a millstone and cut lines in it in order to improve the stone's grinding capacity.

DRESSING MILL STONES: Sharpening and leveling the stones with metal picks and chisels.

ELEVATOR: An endless belt with cups used to transport grain or flour in the mill.

FLAVIN: A water-soluble yellow pigment found in plant and animal tissue.

FLUME: An artificial channel used to convey water from the milldam to a mill; may be a wooden trough or pipe.

FLUTTER WHEEL: An undershot wheel with a series of long paddles connected to arms radiating from a shaft. Requires a substantial water supply, with a swift stream.

FOOLSCAP: a size of paper formerly standard in Great Britain; broadly, a piece of writing paper

FULLING: The process whereby woolen cloth is cleaned, shrunken, and felted to give it the desired texture and consistency.

GRIST: The quantity of grain taken to a mill by a farmer to be ground.

GRISTMILL: A mill used to grind grain for flour or animal food using a waterwheel and millstones.

HEMP MILL: Used to make coarse cloth and cordage from the woody stalk of the hemp plant.

MERCHANT MILL: A mill that purchases and grinds grain into flour or grain products for commercial use.

MIDDLINGS: Coarse bits of the floury part of the wheat grain, small bits of bran may still be present, used for animal feed.

MILLRACE: A ditch, trough, or pipe made of earth, wood, or iron that conveys the water from the dam to the mill.

MILLROOM: An office in the old gristmill.

MILLSTONES: Stones used in grinding grain.

MILLWRIGHT: A craftsman with a knowledge of mills and mill equipment; its construction and installation.

OIL MILL: Mill where flaxseed was crushed to make linseed oil.

OVERSHOT WHEEL: The water reaches the buckets at the top of the wheel by means of a flume. The weight of the water turns the wheel forward.

POWDER MILL: Was used for grinding ingredients of cornpowder.

PURIFIER: Machinery used to separate flour from middlings by passing a current of air through a moving sieve to assist in the separation.

QUERN: A primitive hand mill used for grinding.

SIEVE: To separate the coarse bran from flour. Hand sieves were used in early gristmills to sift corn meal and take the hulls out of buckwheat flour.

SIFTER: A bolter, especially one in which horizontal sieves are used, as distinguished from a centrifuge.

STILL: A distillery.

STONES (burrs, buhrs): A set of millstones used to grind grain.

RUNNER STONE: The upper (rotating) stone or buhr of a pair of millstones.

TAILINGS: The material that is left after the flour is sifted through the bolter or sieve.

TAILRACE: The channel for water to flow back to the stream after passing through the waterwheel or turbine.

TEMPERING: Wetting the grain to toughen the bran and mellow the floury part of the grain.

TRUNK: A wooden trough to carry the water to the waterwheel.

TURBINE: A horizontal waterwheel with vanes; the water runs through the vanes, causing the wheel to turn. Has greater power and efficiency than the waterwheel.

UNDERSHOT WHEEL: The water reaches the buckets near the bottom of the wheel. Only the impact of the water turns the waterwheel, and it moves backward. The undershot wheel is used where the fall of water is insufficient for the more efficient overshot or breast wheel.

WATER HOUSE: A structure that holds the water around the water wheel or turbine.

WATERWHEEL: A hydro-power system; a system for power from a flow of water. Most common use of the waterwheel was to mill flour; other uses were foundry work and pounding linen for use in paper. The wheel is large and typically made of wood, with a number of blades/buckets that are arranged on the outside rim forming the driving surface. The wheel is mounted vertically on a horizontal axle. Waterwheels came in two basic forms: undershot and overshot. Sometimes the early waterwheels were enclosed inside a building or under a roof for protection from snow and ice.

# The Bradley Mill

About 20 years ago, I had been working on a tobacco exhibit for the Agricultural Museum in the old Eastern Market when the Motter Building became available for a larger project.

Warren Smith headed the committee for this larger museum, known today as the York County Heritage Trust's Agricultural and Industrial Museum. I was asked if I could design and build a gristmill to tie agriculture to manufacturing since York County was a leader for decades in both industries.

I was given whatever space I would need, and I chose the wall and corner in the high bay now called the Bradley Building. The request was for a wheel and set of stones to be ready for a formal opening one year later, if not sooner.

The first project would be to find the beams, stones, gears, and wheel, all necessary basic components.

It would not be very interesting to box in the mill with walls, so I decided to build a mill that would be open for all to see and show the construction of the joints and flooring and the machinery drives. The committee agreed to the plan.

Now to put together a team: Since I owned a machinery manufacturing company, Bradley Lifting Corp., that solved this part. I then hired a first-class carpenter who could handle the work.

Leister's Mill in Maryland had to be pulled down. From this mill, a nice set of stones was obtained with pulleys and gearing. The main beam and interlocking floor beams were also salvaged. The 16-foot-diameter Fitz wheel was shot, but the axle and bearings were salvaged, as well as parts for the bag hoist.

A Fitz wheel was designed and constructed in the Bradley fabricating shop. It was erected on the axle and bearings over a trough measuring 18 feet by 6 feet filled with water to a depth of 18 inches. An electric pump lifts the water to the flume, which puts the water on top of the wheel. The wheel then returns the water to the trough.

Gearing drives all motions of the mill, powered by the wheel. A stone speed indicator from a mill in Mechanicsburg and a stone level from Leister's Mill were obtained, both rare items.

Stones were also obtained from a mill near Grand View Golf Course as well as the wood crane. Stones are left open to show the faces ready for dressing. The bag hoist was made from miscellaneous parts from various York mills. The bucket elevator feeds grain into the overhead hopper with a peg screw fed horizontally to the stones. A Lieberknecht Mill was obtained, driven by an electric motor for the actual grinding of corn meal if desired.

A working mill is thus presented for present and future generations to observe, learn from, and enjoy.

Harvey Bradley

# Index